Congressional Research Service

V-22 Osprey Tilt-Rotor Aircraft: Background and Issues for Congress

Jeremiah Gertler
Specialist in Military Aviation

March 10, 2011

Congressional Research Service

7-5700

www.crs.gov

RL31384

CRS Report for Congress ——————————————

Prepared for Members and Committees of Congress

Summary

The V-22 Osprey is a tilt-rotor aircraft that takes off and lands vertically like a helicopter and flies forward like an airplane. Department of Defense plans call for procuring a total of 458 V-22s, including 360 MV-22s for the Marine Corps; 50 CV-22 special operations variants for U.S. Special Operations Command, or USSOCOM (funded jointly by the Air Force and USSOCOM); and 48 HV-22s for the Navy.

Through FY2010, a total of 216 V-22s have been procured—185 MV-22s for the Marine Corps, and 31 CV-22s for USSOCOM. These totals include several V-22s that have been procured in recent years through supplemental appropriations bills. V-22s are currently procured under a $10.4 billion, multiyear procurement arrangement covering the period FY2008-FY2012. The contract, which was awarded on March 28, 2008, covers the procurement of 167 aircraft—141 MV-22s and 26 CV-22s.

The proposed FY2012 budget requested about $2.5 billion in procurement and advance procurement funding for 30 MV-22s, and about $499.9 million in procurement and advance procurement funding for six CV-22s, with $70 million of that from Overseas Contingency Operations funds.

For FY2012, the V-22 program poses a number of potential oversight issues for Congress, including the aircraft's readiness rates, reliability and maintainability, and operational suitability.

FY2011 defense authorization bill: As passed, H.R. 6523, the Ike Skelton National Defense Authorization Act For Fiscal Year 2011 (P.L. 111-383, signed January 7, 2011), did not include program-level detail, so no amount is specified solely for the V-22.

FY2011 DOD appropriations bill: In lieu of a defense appropriations bill, Congress approved a resolution continuing FY2010 spending levels.

On February 15, 2011, the House voted 326 to 105 against an amendment to a proposed continuing resolution for FY2011 that would have reduced V-22 funding by $415 million.

Contents

Figures

Tables

Appendixes

Contacts

Introduction

The V-22 Osprey is a tilt-rotor aircraft that takes off and lands vertically like a helicopter and flies forward like an airplane. Department of Defense (DOD) plans call for procuring a total of 458 V-22s—360 MV-22s for the Marine Corps; 50 CV-22 special operations variants for U.S. Special Operations Command, or USSOCOM (funded jointly by the Air Force and USSOCOM); and 48 HV-22s for the Navy.

Through FY2010, a total of 216 V-22s have been procured—185 MV-22s for the Marine Corps, and 31 CV-22s for USSOCOM. V-22s are currently procured under a $10.4 billion, multiyear procurement (MYP) arrangement covering the period FY2008-FY2012. The MYP contract, which was awarded on March 28, 2008, covers the procurement of 167 aircraft—141 MV-22s and 26 CV-22s.

The proposed FY2012 budget requested funding for the procurement of 30 MV-22s and 7 CV-22s, with one CV-22 to be paid for from Overseas Contingency Operations (OCO) funding. The budget requested about $2.5 billion in procurement and advance procurement funding for procurement of MV-22s, and about $499.9 million in procurement and advance procurement funding for procurement of CV-22s, with $70 million of that from OCO.

For FY2012, the V-22 program poses potential a number of potential oversight issues for Congress, including the aircraft's readiness rates, reliability and maintainability, and operational suitability.

Recent Developments

Proposal to Eliminate V-22 funding in FY2011

On February 15, 2011, the House voted 326 to 105 against a proposed cut to the FY2011 V-22 budget.[1]

The amendment proposed to eliminate FY2011 funding for the V-22 by reducing the amount for Aircraft Procurement, Navy by $22 million and for Aircraft Procurement, Air Force by $393 million.

Proponents cited cost overruns and argued that the V-22 did not meet operational requirements, citing the 2009 GAO report. Opponents noted the V-22's performance in Iraq and Afghanistan and highlighted advances made since the aircraft's development period.

Deficit Reduction Commission Recommendation

On December 3, 2010, the National Commission on Fiscal Responsibility and Reform released its report on ways to decrease the United States' national debt. The commission chairs' illustrative suggestions included:

---

[1] H.Amdt. 13 to H.R. 1, Full-Year Continuing Appropriations Act, 2011, roll call vote 43.

End procurement of the V-22 Osprey. The V-22 Osprey was designed to meet the amphibious assault needs of the Marine Corps, the strike rescue needs of the Navy, and the needs of long range special operations forces (SOF) missions of U.S. Special Operations Command. However, the V-22 has had a troubled history with many developmental and maintenance problems, including critical reports by GAO and others.... The proposed change to terminate acquisition of V-22 at 288 aircraft, close to two-thirds of the planned buy, would substitute MH-60 helicopters to meet missions that require less range and speed, and could save $1.1 billion in 2015.[2]

Background

The V-22 In Brief

The V-22 Osprey is a tilt-rotor aircraft that takes off and lands vertically like a helicopter and flies forward like an airplane. For taking off and landing, the aircraft's two wingtip-mounted engine nacelles are tilted upward, so that the rotors function like a helicopter's rotor blades. For forward flight, the nacelles are rotated 90 degrees forward, so that the rotors function like an airplane's propellers. The Navy states that the V-22 "performs VTOL [vertical takeoff and landing] missions as effectively as a conventional helicopter while also having the long-range cruise abilities of a twin turboprop aircraft."[3]

The MV-22 is designed to transport 24 fully equipped Marines at a cruising speed of about 250 knots (about 288 mph), exceeding the performance of the Marine Corps CH-46 medium-lift assault helicopters that MV-22s are to replace. The CV-22 has about 90% airframe commonality with the MV-22; the primary differences between the two variants are in their avionics. The CV-22 is designed to carry 18 troops, with auxiliary fuel tanks increasing the aircraft's combat radius to about 500 miles.

Figure 1 shows a picture of an MV-22 with its engine nacelles rotated at about a 45-degree angle, or roughly halfway between the upward VTOL position and the forward-flight position.

---

[2] National Commission on Fiscal Responsibility and Reform draft document, "$200 Billion in Illustrative Savings," November 12, 2010.

[3] U.S. Navy Fact File, "V-22A Osprey tilt rotor aircraft," available at http://www navy mil/navydata/fact_display.asp? cid=1200&tid=800&ct=1&page=1.

Figure 1. MV-22 Osprey

Source: Military-Today.com: http://www.military-today.com/helicopters/bellboeing_v_22_osprey.jpg.

Intended Missions

The V-22 is a joint-service, multi-mission aircraft. The Navy, which is the lead service for the V-22 program, states that "the Marine Corps version, the MV-22A, will be an assault transport for troops, equipment and supplies, and will be capable of operating from ships or from expeditionary airfields ashore. The Navy's HV-22A will provide combat search and rescue, [as well as] delivery and retrieval of special warfare teams along with fleet logistic support transport. The Air Force CV-22A will conduct long-range special operations missions."[4] Specific CV-22 missions include "long range, high speed infiltration, exfiltration, and resupply to Special Forces teams in hostile, denied, and politically sensitive areas."[5]

Marine Corps leaders believe that the MV-22 provides significant operational advantages compared to the CH-46, particularly in terms of speed in forward flight. The V-22 has been the Marine Corps' top aviation priority for many years.[6]

Regarding the V-22's role as a combat search and rescue aircraft, particularly as a possible replacement for a canceled CSAR helicopter program called CSAR-X, an October 9, 2009, press report stated:

---

[4] U.S. Navy Fact File, "V-22A Osprey tilt rotor aircraft," available at http://www navy mil/navydata/fact_display.asp?cid=1200&tid=800&ct=1&page=1.

[5] United States Special Operations Command, Fiscal Year (FY) 2009 Budget Estimates, February 2008, Procurement, Defense-Wide, Exhibit P-40 Budget Item Justification Sheet, page 1 of 13 (overall document page 59 of 192).

[6] See, for example, Department of the Navy, *Highlights of the Department of the Navy FY 2010 Budget*, May 2009, p. 5-11.

Boeing officials last week insisted that their V-22 Osprey is a viable aircraft for risky combat search-and-rescue missions despite findings in a recent Pentagon report claiming the tiltrotor was outclassed in the rescue mission by other special operations helicopters.

"We still see [the Osprey] as very effective" in the CSAR role, said Gene Cunningham, Boeing's V-22 program manager, during an Oct. 2 telephone interview. "I think, in a CSAR configuration, the aircraft fulfills all of the requirements" for the mission....

Meanwhile, the Air Force and Marines are working to boost the V-22's firepower with the addition of a removable, belly-mounted 360-degree minigun linked to a sensor package that will give the crew chief a complete view of the outside environment. A limited number of the BAE-built weapon is set to deploy with Marine Corps Ospreys to Afghanistan this fall.

Cunningham also noted the aircraft's impressive speed—277 miles per hour in cruise mode—as giving it an advantage in the rescue role.[7]

Key Contractors

The V-22 was developed and is being produced by Bell Helicopter Textron of Fort Worth, TX, and Boeing Helicopters of Philadelphia, PA. The aircraft's engines are produced by Allison Engine Company of Indianapolis, IN, a subsidiary of Rolls-Royce North America. Fuselage assembly is performed in Philadelphia, PA. Drive system rotors and composite assembly is performed in Fort Worth, TX, and final assembly and delivery is performed in Amarillo, TX.

Procurement Quantities

Total Quantities

Department of Defense (DOD) plans call for procuring a total of 458 V-22s—360 MV-22s for the Marine Corps; 50 CV-22 special operations variants for U.S. Special Operations Command, or USSOCOM (funded jointly by the Air Force and USSOCOM); and 48 HV-22s for the Navy.[8]

Through FY2010, a total of 216 V-22s have been procured—185 MV-22s for the Marine Corps, and 31 CV-22s for USSOCOM. These totals include several V-22s that have been procured in recent years through supplemental appropriations bills. No HV-22s have yet been procured for the Navy.

---

[7] John Reed, "After Negative Report, Boeing Defends V-22's Ability To Fly CSAR Mission," *Inside the Air Force*, October 9, 2009. See also John Reed, "JFCOM Rescue Study Finds V-22 Ospreys Ill Suited For CSAR Role (Updated)," *Inside the Air Force*, October 2, 2009.

[8] Like some other tactical aviation, the total number of V-22 aircraft planned for procurement has decreased over time. In 1989 the Defense Department projected a 663-aircraft program with six prototypes and 657 production aircraft (552 MV-22s, 55 CV-22s, and 50 HV-22s). As projected in 1994, however, the program comprised 523 production aircraft (425 MV-22s, 50 CV-22s, and 48 HV-22s). The Quadrennial Defense Review (QDR), released May 19, 1997, recommended accelerated procurement of 458 production aircraft.

Annual Quantities

Table 1 shows annual procurement quantities of MV-22s and CV-22s funded through DOD's regular (aka "base") budget. The table *excludes* the several V-22s that have been procured in recent years through wartime supplemental appropriations bills as replacements for legacy helicopters lost as a result of wartime operations.

Table 1. Annual V-22 Procurement Quantities

(Excludes V-22s procured through wartime supplemental funding)

FY	MV-22	CV-22	Total
1997	5	0	5
1998	7	0	7
1999	7	0	7
2000	11	0	11
2001	9	0	9
2002	9	0	9
2003	11	0	11
2004	9	2	11
2005	8	3	11
2006	9	2	11
2007	14	2	16
2008	19	5	24
2009	30	6	36
2010	30	5	35
2011	30	5	35
2012 *(requested)*	30	5	35

Source: Prepared by CRS based on DOD data.

Notes: Figures shown exclude several additional V-22s procured in recent years (and one CV-22 requested in 2012) with wartime supplemental funding.

Multiyear Procurement (MYP) for FY2008-FY2012

V-22s are currently being procured under a $10.4-billion, multiyear procurement (MYP) arrangement covering the period FY2008-FY2012. The MYP contract, which was awarded on March 28, 2008, covers the procurement of 167 aircraft—141 MV-22s and 26 CV-22s. DOD expects the multiyear contract to save $427 million when compared to the use of annual contracting.[9]

---

[9] Christopher J. Castelli, "Navy Awards $10.4 Billion V-22 Multiyear Deal," *Inside Washington Publishers*, March 28, 2008, online at http://www.insidedefense.com.

Cost and Funding

Total Program Cost

In December 2009, its most recent report, DOD estimated the total acquisition cost of a 458-aircraft V-22 program at about $52.9 billion in base year 2005 dollars, including about $10.1 billion for research and development, about $42.7 billion for procurement, and $113.2 million for military construction (MilCon). The program was estimated to have a program acquisition unit cost, or PAUC (which is total acquisition cost divided by the number of aircraft), of about $109 million and an average procurement unit cost, or APUC (which is procurement cost divided by the number of aircraft), of about $83.7 million.[10]

Prior-Year Funding

In then-year dollars, the V-22 program from FY1982 through FY2009 received a total of about $29.1 billion in funding, including about $9.6 billion for research and development, about $19.4 billion for procurement, and about $72.9 million for MilCon.[11] These figures *exclude* wartime supplemental funding that has been provided in addition to DOD's regular (aka "base") budget. As mentioned earlier, this supplemental funding has, among other things, funded the procurement of several V-22s.

FY2012 Funding Request

The proposed FY2012 budget requests funding for the procurement of 30 MV-22s and 6 CV-22s, with one CV-22 to be paid for from Overseas Contingency Operations (OCO) funding. The budget requests about $2.5 billion in procurement and advance procurement funding for procurement of MV-22s, and about $499.9 million in procurement and advance procurement funding for procurement of CV-22s, with $70 million of that from OCO.

Program History and Milestones

The V-22 program began in the early 1980s.[12] The aircraft experienced a number of development challenges relating to affordability, safety, and program management. Crashes of prototypes occurred in June 1991 (no fatalities) and July 1992 (seven fatalities). Two additional crashes occurred in April 2000 (19 fatalities) and December 2000 (4 fatalities). The V-22's development challenges were a topic of considerable oversight and debate during the 1990s.

---

[10] *Selected Acquisition Report[:] V-22*, DOD, December 31, 2009.

[11] Ibid.

[12] The V-22 is based on the XV-15 tilt-rotor prototype which was developed by Bell Helicopter and first flown in 1977. The Department of Defense began the V-22 program first under Army leadership; the Navy and Marine Corps subsequently assumed leadership. The V-22 program was given Milestone 0 approval in December 1981 as the Joint Services Aircraft program, and Milestone I approval in December 1982, at which time the program's acquisition strategy was approved. A preliminary design contract for the aircraft was awarded in April 1983 to a Bell-Boeing industry team, which was the only competitor for the program. The aircraft was designated the V-22 Osprey in January 1985. The program was given Milestone II approval in April 1986, initiating system development and demonstration. A full-scale development (FSD) contract was awarded in May 1986.

The acquisition program baseline (APB) for the V-22 has been revised numerous times over the program's history. The V-22 program has undergone restructuring to accommodate recommendations from outside experts and DOD managers.

The George H. W. Bush Administration proposed terminating the V-22 program in 1989 as part of its proposed FY1990 budget, and continued to seek the cancellation of the program through 1992. Congress rejected these proposals and kept the V-22 program alive. The Marine Corps' strong support for the program was reportedly a key reason for Congress's decision to keep the program going.

The MV-22 achieved Initial Operational Capability (IOC) in June 2007. The CV-22 achieved IOC in March 2009.[13]

For additional discussion of the history of the V-22 program, see **Appendix B**.

Initial Deployments

The first deployment of MV-22s began in September 2007, with the deployment of 10 MV-22s to Al Anbar province in Iraq.[14] The Marine Corps has lauded the extended range, speed, and payload that the Osprey possesses in comparison to helicopters it is intended to replace as instrumental to the success of time-critical interdiction and medical evacuation missions during the deployment.[15]

The first deployment of CV-22s, which involved four aircraft sent to Mali, occurred in December 2008. The aircraft participated in a multinational exercise. Those involved in the deployment report successfully self-deploying the squadron to a remote and austere location and conducting simulated long-range, air-drop, and extraction missions.[16]

MV-22s arrived in Afghanistan in November 2009[17] and continue to operate there today. The MV-22s used in Afghanistan have added armament; a 7.62-mm gun is mounted below the belly of the

---

[13] In August 1995, the V-22 contract was modified to include the CV-22 as a special operations version of the aircraft. The CV-22 completed CDR in December 1998. CV-22 flight testing began in February 2000 and was completed in October 2007. A production contract for long lead items for the CV-22 was awarded in June 2000. CV-22 Initial Operational Test and Evaluation (IOT&E) began in June 2006.

[14] The first MV-22 prototype flow in helicopter mode in March 1989. The first forward-facing flight occurred in September 1989. The MV-22 completed Critical Design Review (CDR) in December 1994. The first low-rate initial production (LRIP) contract was awarded in June 1996, and the first delivery of an LRIP aircraft occurred in May 1999. Technical evaluation (TECHEVAL) began in July 1999 and was completed in September 1999. Operational evaluation (OPEVAL) began in November 1999 and was completed in July 2000.

In January 2001, an MV-22 squadron commander was relieved of duty after admitting to falsifying maintenance records, and three Marines were found guilty of misconduct in September 2001. In April 2001, a blue ribbon panel formed by Secretary of Defense William Cohen recommended continuing with the V-22 program in restructured form.

Phase II of the MV-22's OPEVAL began in March 2005 and was completed in June 2005. The program was given Milestone III approval, permitting full-rate production, in October 2005.

[15] Michael Fabey, "Ospreys Proving Mettle in Counter-IED, Medevac Missions," *Aerospace Daily & Defense Report*, January 31, 2008, p. 4.

[16] 1st Lt. Lauren Johnson. "CV-22s Complete First Operational Deployment." *Air Force News.* December 3, 2008.

[17] Jay Price, "Controversial 'Osprey' makes combat debut in Afghanistan," *McClatchy News Service*, December 5, 2009.

---

aircraft, and the standard rear-mounted 7.62 is replaced by a .50-caliber gun. CV-22s arrived in early 2010. One CV-22 crashed in April 2010, with the loss of four lives.[18]

Foreign Military Sales

To date, there have been no sales of the V-22 to foreign military forces. "[T]he industry team is also in talks with several countries about potential V-22 sales, including the UK, Japan and Israel. Bell and Boeing have already responded to Canada's request for information for a new fixed-wing search and rescue aircraft."[19] Other unnamed nations were reported to have expressed interest following a 2010 "embassy day" flight demonstration.[20]

GAO Assessments

March 2010 GAO Report

A March 2010 Government Accountability Office (GAO) report on the V-22 program stated:

Technology Maturity

Although the program office considers V-22 critical technologies to be mature and its design stable, the program continues to correct deficiencies and make improvements to the aircraft. For example, the engine air particle separator (EAPS), which keeps debris out of the engines, and has been tied to a number of engine fires caused by leaking hydraulic fluids contacting hot engine parts. Previous design changes did not fully correct this problem or other EAPS problems. According to program officials a root cause analysis is underway and they are exploring ways to improve reliability and safety of EAPS. Further, they believe that improved EAPS performance could reduce EAPS shutdowns and help to extend engine service life beyond its current average of 600 hours. According to program officials the program has purchased eight belly mounted all quadrant (360 degrees) interim defensive weapon system mission kits. Five kits are currently on deployed V-22 aircraft. The aircraft has a key performance parameter (KPP) requirement to carry 24 combat equipped troops. The MV-22's shipboard pre-deployment exercise found that planning for fewer troops is needed to allow for additional space for equipment, including larger personal protective equipment. When retracted, the belly-mounted gun would reduce internal space and it will not meet the KPP of 24 combat equipped troops. According to program officials, incremental upgrades to the IPS are being fielded in concert with an overall strategy to improve IPS reliability. These incremental upgrades are now being fielded on some deployed aircraft, including the V-22s attached to the squadron deployed to Afghanistan, where icing conditions are more likely to be encountered. The program expects to make additional improvements to the IPS which could require retrofits to existing aircraft.

Production Maturity

The V-22 is in the third year of a 5-year contract for 167 aircraft. According to the program office, the production rate will be 35 aircraft per year for fiscal years 2010 through 2012. The

---

[18] See, inter alia, Bob Cox, "Findings on Osprey crash in Afghanistan overturned," *Fort Worth Star-Telegram*, December 16, 2010.

[19] Caitlin Harrington, "US Expected To Extend Contract For V-22 Osprey," *Jane's Defence Weekly*, June 2, 2010.

[20] Bettina H. Chavanne, "Foreign Interest In V-22 Ramps Up, Program Office Says," *Aerospace Daily*, June 25, 2010.

program is planning and budgeting for cost savings that would result from a second multiyear procurement contract that would begin in fiscal year 2013.

Other Program Issues

The MV-22's shipboard pre-deployment training revealed challenges related to required aircraft maintenance and operations. Due to the aircraft's design, many components of the aircraft are inaccessible until the aircraft is towed from its parking spot. Shipboard operations were adjusted to provide 24 hour aircraft movement capability. Temporary work-arounds were also identified to mitigate competition for hangar deck space, as well as to address deck heating issues on smaller ships caused by the V-22's exhaust. Operational restrictions were also in place that required one open spot between an MV-22 when landing or taking off and smaller aircraft to avoid excessive buffeting of the lighter helicopters caused by the downwash of the Osprey. According to program officials, another restriction that limited takeoffs and landings from two spots on LHD-class ships has since been corrected with the installation of a new flight control software upgrade. Despite the restrictions, the amphibious assault mission was concluded with half the total number of aircraft, in less time, and over twice the distance compared to conducting the mission using traditional aircraft. However, the speed, altitude, and range advantages of the MV-22 will require the Marine Corps to reevaluate escort and close air support tactics and procedures. According to the program office, during the first sea deployment in 2009, the MV-22 achieved a mission capable rate of 66.7 percent. This still falls short of the minimum acceptable (threshold) rate of 82 percent. The mission capable rate achieved during three Iraq deployments was 62 percent average. The program is also taking various steps to improve the system's overall operational availability and cost to operate by addressing premature failure of selected components and establishing a steering committee to analyze factors that affect readiness and impact operations and support costs.[21]

May 2009 GAO Report

A May 2009 Government Accountability Office (GAO) report on the V-22 program stated:

> As of January 2009, the 12 MV-22s (Marine Corps variant of the V-22) in Iraq successfully completed all missions assigned in a low threat theater of operations—using their enhanced speed and range to engage in general support missions and deliver personnel and internal cargo faster and farther than the legacy helicopters being replaced. Noted challenges to operational effectiveness raise questions about whether the MV-22 is best suited to accomplish the full repertoire of missions of the helicopters it is intended to replace. Additionally, suitability challenges, such as unreliable component parts and supply chain weaknesses, led to low aircraft availability rates.

> MV-22 operational tests and training exercises identified challenges with the system's ability to operate in other environments. Maneuvering limits and challenges in detecting threats may affect air crew ability to execute correct evasive actions. The aircraft's large size and inventory of repair parts created obstacles to shipboard operations. Identified challenges could limit the ability to conduct worldwide operations in some environments and at high altitudes similar to what might be expected in Afghanistan. Efforts are underway to address these deficiencies, but some are inherent in the V-22's design.

---

[21] Government Accountability Office, *Defense Acquisitions[:] Assessments of Selected Weapon Programs*, GAO-10-388SP, March 2010, p. 132.

V-22 costs have risen sharply above initial projections—1986 estimates (stated in fiscal year 2009 dollars) that the program would build nearly 1000 aircraft in 10 years at $37.7 million each have shifted to fewer than 500 aircraft at $93.4 million each—a procurement unit cost increase of 148 percent. Research, development, testing, and evaluation costs increased over 200 percent. To complete the procurement, the program plans to request approximately $25 billion (in then-year dollars) for aircraft procurement. As for operations and support costs (O&S), the Marine Corps' V-22's cost per flight hour today is over $11,000—more than double the targeted estimate.[22]

Issues for Congress

Readiness Rates

Readiness rates for both the CV-22 and MV-22 are lower than those for more traditional aircraft. Overall,

> [t]he V-22, in its most recent testing to evaluate upgrades, was available only 57 percent of the time it was required to fly, rather than the specification of 82 percent, said a new report by Michael Gilmore, the Pentagon's director of operational test and evaluation.[23]

The testing report identified the main issue as "unreliable parts," noting that

> [w]hen the V-22 was flying ... it "met or exceeded" all but one reliability and maintenance requirement, proving effective "in a wide range of approved high-altitude scenarios" in Marine Corps operations.[24]

The FY2010 mission-capable rate for the Air Force CV-22 fleet was reported as 54.3%.

> No common problem such as a software glitch or engine malfunction led to the Osprey's low rate, said Col. Peter Robichaux, who oversees the health of Air Force Special Operations Command aircraft.
>
> For Robichaux, the Osprey's low rate is a statistical quirk—not an indicator of its long-term viability.
>
> "The numbers are a result of our small fleet size," said Robichaux..."That can drive the numbers down." The Air Force has 16 CV-22s.... Taking one plane off the flight schedule for a day pushes down the mission-capable rate for that day by about 6 percentage points.[25]

The Marine Corps MV-22 has maintained a higher readiness rate, with deployed Ospreys "in the low 70th percentile."[26]

---

[22] Government Accountability Office, *Defense Acquisitions[:] Assessments Needed to Address V-22 Aircraft Operational and Cost Concerns to Define Future Investments*, GAO 09-482, May 2009, summary page.

[23] Tony Capaccio, "Pentagon: Unreliable Parts Still Plague V-22," *Philadelphia Inquirer*, January 13, 2011.

[24] Ibid.

[25] Bruce Rolfsen, "U.S. Air Force Ospreys Ready Just Half the Time," *Defense News*, November 15, 2010.

[26] Emelie Rutherford, "Trautman: F-35 Remains Primary Issue For Successor ," *Defense Daily*, September 28, 2010.

[Marine Assistant Commandant for Aviation George] Trautman said the service's other assault-support airplanes had readiness rates in the low 70s. And he said it can't be overlooked that Afghanistan has "the most harsh air environment in the world," because it is filled with fine talcum-powder-like dust that is very hard on airplanes.

"We've been able to maintain those readiness rates by sparing the airplane out, by putting spare parts in place at a higher rate than we would like," Trautman said. "The global readiness of the V-22 is something that still concerns both Bell Boeing and myself as we work through some of the supplier issues and some of the other reliability issues that haven't turned out to be exactly as the engineers predicted several years ago. But we're holding our own."[27]

A related oversight issue for Congress concerns the reliability and maintainability of in-service V-22s, factors that bear directly on readiness.

May 2009 Navy and Marine Corps Testimony

At a May 19, 2009, hearing on Navy and Marine Corps aviation procurement programs before the Seapower and Expeditionary Forces Subcommittee of the House Armed Services Committee, Navy and Marine Corps officials testified that:

As we continue to explore the tremendous capabilities of tilt-rotor aircraft and look forward to employing Osprey both aboard ship and in new theaters of operation, we are learning valuable lessons with respect to reliability and maintainability. Like other types of aircraft in the early operational phase of their lifecycles, the MV-22 has experienced lower-than-desired reliability of some components and therefore higher operations and support costs. With the cooperation and support of our industry partners, we are tackling these issues head on, with aggressive logistics and support plans that will increase the durability and availability of the parts needed to raise reliability and concurrently lower operating costs of this aircraft.[28]

May 2009 GAO Report

The May 2009 GAO report on the V-22 program cited earlier stated the following regarding the aircraft's reliability and maintainability:

Availability challenges continue to affect the MV-22. In Iraq, the V-22's mission capability (MC) and full mission capability (FMC) rates fell significantly below required levels and significantly below rates achieved by legacy helicopters. The MV-22 has a stated MC threshold (minimum acceptable) requirement of 82 percent and an objective (desired) of 87 percent. In Iraq, the three MV-22 squadrons averaged mission capability rates of about 68, 57, and 61 percent respectively. This experience is not unique to the Iraq deployment, as low MC rates were experienced for all MV-22 squadrons, in and out of Iraq. The program has modified the MC requirement by stating that this threshold should be achieved by the time

---

[27] Ibid.

[28] Statement of Vice Admiral David Architzel, USN, Principal Military Deputy, Research, Development and Acquisition, LTGEN George J. Trautman III, USMC, Deputy Commandant for Aviation, [and] RADM Allen G. Myers, USN, Director of Warfare Integration, before the Seapower and Expeditionary Warfare [sic: Forces] Subcommittee of the House Armed Services Committee [hearing] on [The] Department of the Navy's Aviation Procurement Program, May 19, 2009, pp. 7-8.

the fleet completes 60,000 flight hours, which officials expect to occur sometime near the end of 2009. Figure 4 illustrates the MC rates between October 2006 and October 2008.

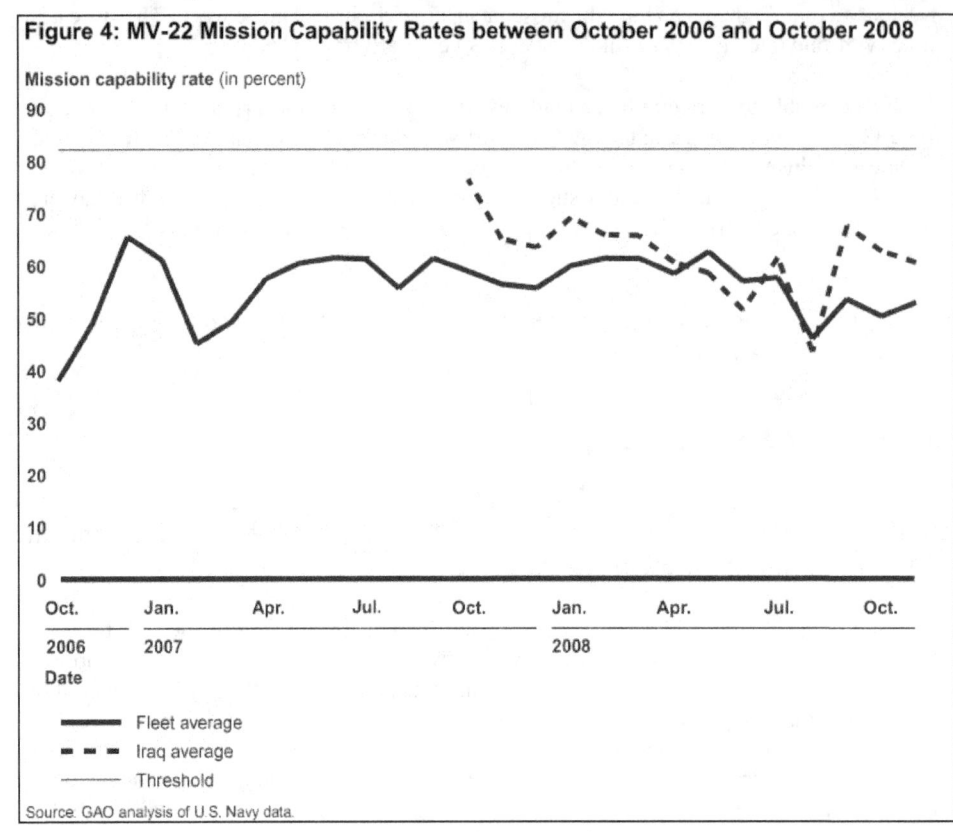

Figure 4: MV-22 Mission Capability Rates between October 2006 and October 2008

Source: GAO analysis of U.S. Navy data.

By comparison, the mission capability rates of the Iraq-based CH-46Es and CH-53s averaged 85 percent or greater during the period of October 2007 to June 2008.

Although FMC is no longer a formal requirement, it continues to be tracked as an indicator of aircraft availability. The Osprey's FMC rate of 6 percent in Iraq from October 2007 to April 2008 was significantly short of the 75 percent minimum requirement established at the program's outset. According to MV-22 officers and maintainers, the low FMC rate realized was due in part to unreliability of V-22 Ice Protection System (IPS) components. Although the faulty IPS had no effect on the MV-22's ability to achieve missions assigned in Iraq, in other areas, where icing conditions are more likely to be experienced—such as Afghanistan—IPS unreliability may threaten mission accomplishment.

Although MV-22 maintenance squadrons stocked three times as many parts in Iraq as the number of deployed MV-22 aircraft called for, they faced reliability and maintainability challenges. Challenges were caused mostly by an immature parts supply chain and a small number of unreliable aircraft parts, some of which have lasted only a fraction of their projected service life.

The MV-22 squadrons in Iraq made over 50 percent more supply-driven maintenance requests than the average Marine aviation squadron in Iraq. A lack of specific repair parts was a problem faced throughout the Iraq deployments despite deploying with an inventory of spare parts to support 36 aircraft, rather than the 12 MV-22 aircraft actually deployed. Despite the preponderance of parts brought to support the MV-22s in Iraq, only about 13 percent of those parts were actually used in the first deployment. In addition, some aircraft components wore out much more quickly in Iraq than expected, which led to shortages. Thirteen MV-22 components

accounted for over half the spare parts that were not available on base in Iraq when requested. Those components lasted, on average, less than 30 percent of their expected life, with six lasting less than 10 percent of their expected life. The shortages caused MV-22 maintainers to cannibalize parts from other MV-22s to keep aircraft flying, and significantly increased maintenance hours. Parts were cannibalized not only from MV-22s in Iraq but also from MV-22s in the United States and from the V-22 production line. The shortages also contributed to the low mission capability rates, as an aircraft in need of maintenance or spare parts may not be considered mission capable. Figure 5 depicts both the percentage of predicted mean flight hours before failure achieved by these 13 parts and their average requisition waiting time during the Iraq deployments.

Figure 5: Attained Percentage of Predicted Mean Flight Hours before Failure and Requisition Wait Time for Top 13 Parts Degrading MV-22 Mission Capability

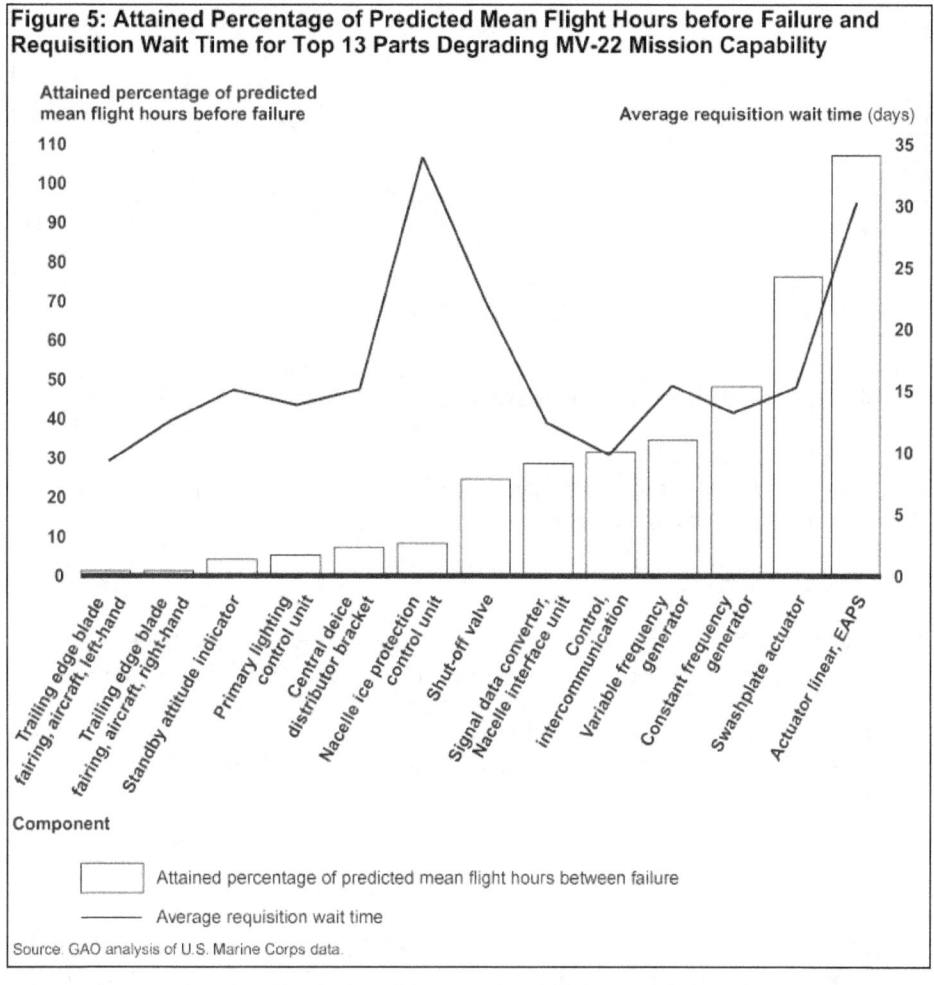

Source: GAO analysis of U.S. Marine Corps data.

The engines on the MV-22s deployed in Iraq also fell short of their estimated "on-wing" service life, lasting less than 400 hours before having to be replaced. The program estimated life is 500-600 hours. The program office noted that there is no contractually documented anticipated engine service life. Figure 6 illustrates the average engine time on wing for the three MV-22 squadrons that have been deployed to Iraq.

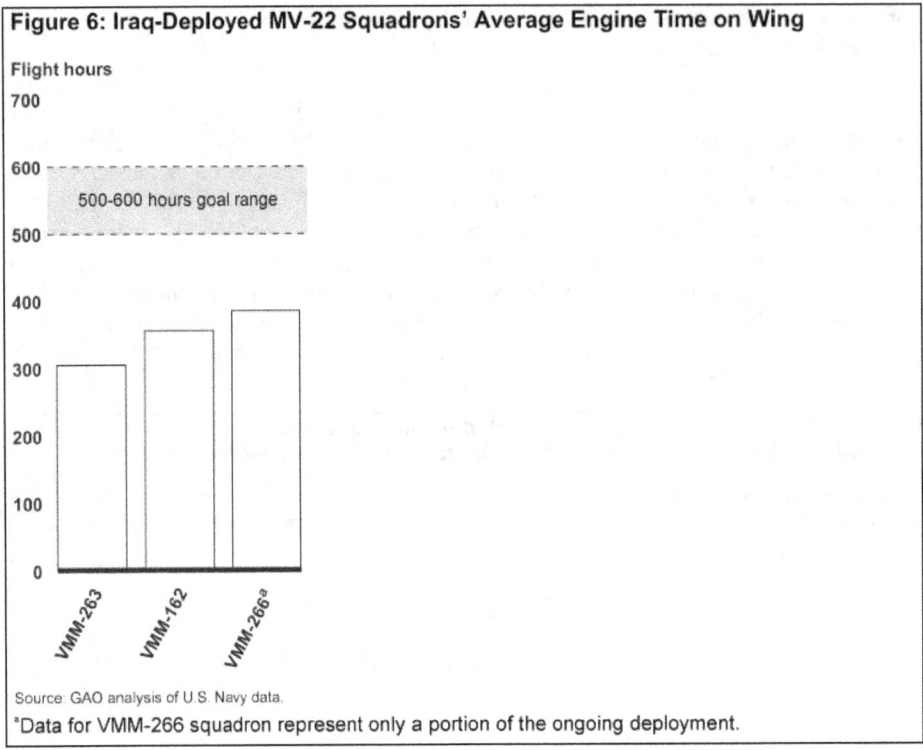

Figure 6: Iraq-Deployed MV-22 Squadrons' Average Engine Time on Wing

Source: GAO analysis of U.S. Navy data.
ªData for VMM-266 squadron represent only a portion of the ongoing deployment.

Squadron maintainers explained that the lower engine life span has not affected aircraft availability, as spare engines are readily available and easily replaced. Program officials plan to replace the existing power-by-the-hour engine sustainment contract with Rolls Royce, which expires in December 2009, with a new sustainment contract.17 According to the program office, the new engine sustainment contract is likely to result in higher engine support costs—an issue further discussed later in this report.[29]

Subsequent to the GAO report,

> The U.S. Marine Corps says MV-22 performance and reliability are improving, but operators are still pushing for further enhancements, including improving the system's firepower.... Service officials say they are also benefiting from reliability improvements now being introduced. The engine air particle accelerator, for instance, has been upgraded to have less failures and do a better job filtering sand. Blades have also been upgraded, as have swashplate actuators.[30]

Operational Capabilities

Another potential oversight issue for Congress for the V-22 program concerns the degree to which the V-22 has demonstrated certain operational capabilities. The May 2009 GAO report cited earlier states:

---

[29] Government Accountability Office, *Defense Acquisitions[:] Assessments Needed to Address V-22 Aircraft Operational and Cost Concerns to Define Future Investments*, GAO 09-482, May 2009, pp. 14-18.

[30] Robert Wall, "U.S. Marines See MV-22 Improvements, Want More," *Aerospace Daily*, June 24, 2010.

As of January 2009, the 12 MV-22s stationed in Iraq had successfully completed all missions assigned to them in what is considered an established, low-threat theater of operations. The deployments confirmed that the V-22's enhanced speed and range enable personnel and internal cargo to be transported faster and farther than is possible with the legacy helicopters it is replacing. The aircraft also participated in a few AeroScout missions and carried a limited number of external cargo loads. However, questions have arisen as to whether the MV-22 is best suited to accomplish the full mission repertoire of the helicopters it is intended to replace. Some challenges in operational effectiveness have been noted....

The Marine Corps considers the MV-22 deployments in Iraq to have been successful, as the three squadrons consistently fulfilled assigned missions. Those missions were mostly general support missions—moving people and cargo—in the low-threat operational environment that existed in Iraq during their deployments. The aircraft's favorable reviews were based largely on its increased speed and range compared with legacy helicopters. According to MV-22 users and troop commanders, its speed and range "cut the battlefield in half," expanding battlefield coverage with decreased asset utilization and enabling it to do two to three times as much as legacy helicopters could in the same flight time. In addition, the MV-22's ability to fly at higher altitudes in airplane mode enabled it to avoid the threat of small arms fire during its Iraq deployment....

Commanders and operators have noted that the speed and range of the Osprey offered some significant advantages over the legacy platforms it replaced during missions performed in Iraq, including missions that would have been impossible without it. For example, it enabled more rapid delivery of medical care; missions that had previously required an overnight stay to be completed in a single day; and more rapid travel by U.S. military and Iraqi officials to meetings with Iraqi leaders, thus allowing greater time for those meetings.

While in Iraq, the MV-22 also conducted a few AeroScout raid and external lift missions. These types of missions were infrequent, but those that were carried out were successfully completed. Such missions, however, were also effectively carried out by existing helicopters. AeroScout missions are made by a combination of medium-lift aircraft and attack helicopters with a refueling C-130 escort that, according to Marine Corps officers, find suspicious targets and insert Marines as needed to neutralize threats. In participating in these missions, the MV-22 was limited by operating with slower legacy helicopters—thus negating its speed and range advantages. Similarly, external lift missions do not leverage the advantages of the V-22. In fact, most Marine equipment requiring external transport is cleared only for transit at speeds under 150 knots calibrated airspeed (kcas), not the higher speeds at which the MV-22 can travel with internal cargo or passengers. According to Iraq-based MV-22 squadron leadership, the CH-53, which is capable of lifting heavier external loads, was more readily available than the MV-22 to carry out those missions and, as such, was generally called on for those missions, allowing the MV-22 to be used more extensively for missions that exploit its own comparative strengths.

The introduction of the MV-22 into Iraq in combination with existing helicopters has led to some reconsideration of the appropriate role of each. Battlefield commanders and aircraft operators in Iraq identified a need to better understand the role the Osprey should play in fulfilling warfighter needs. They indicated, for example, that the MV-22 may not be best suited for the full range of missions requiring medium lift, because the aircraft's speed cannot be exploited over shorter distances or in transporting external cargo. These concerns were also highlighted in a recent preliminary analysis of the MV-22 by the Center for Naval Analysis, which found that the MV-22 may not be the optimal platform for those missions.

The MV-22's Iraq experience also demonstrated some limitations in situational awareness that challenge operational effectiveness. For example, some MV-22 crew chiefs and troop commanders in Iraq told us that they consider a lack of visibility of activity on the ground from the V-22's troop cabin to be a significant disadvantage—a fact previously noted in operational testing. They noted that the V-22 has only two small windows. In contrast, combat Marines in Iraq stated that the larger troop compartment windows of the CH-53 and CH-46 offer improved ability to view the ground, which can enhance operations. In addition, CH-53s and CH-46s are flown at low altitude in raid operations. According to troop commanders this low altitude approach into the landing zones combined with the larger windows in CH-53s and CH-46s improves their (the troop commanders) situational awareness from the troop compartments, compared with the situational awareness afforded troop commanders in the MV-22s with its smaller windows and use of high altitude fast descent approach into the landing zone. The V-22 program is in the process of incorporating electronic situational awareness devices in the troop cabin to off-set the restricted visibility. This upgrade may not fully address the situational awareness challenges for the crew chief, who provides visual cues to the pilots to assist when landing. Crew chiefs in Iraq agree that the lack of visibility from the troop cabin is the most serious weakness of the MV-22.[31]

Afghan deployment

Reports indicate that commanders are pleased with the performance of V-22s in Afghanistan. However, as operations there are still underway, no comprehensive look has yet been undertaken to compare the Osprey's actual performance to projections and studies. CRS anticipates including such evaluations in future versions of this report.

June 23, 2009, Hearing on V-22 Program

A June 23, 2009, hearing before the House Oversight and Government Reform Committee reviewed a number of issues concerning the V-22 program, including those discussed above.[32]

---

[31] Government Accountability Office, *Defense Acquisitions[:] Assessments Needed to Address V-22 Aircraft Operational and Cost Concerns to Define Future Investments*, GAO 09-482, May 2009, pp. 11-14.

[32] The hearing was originally scheduled for May 21, 2009, but the hearing was adjourned after a few minutes and laer rescheduled for June 23, 2009. The chairman of the committee, Representative Edolphus Towns, stated the following at the opening of the May 21 hearing:

> Good morning. Thank you all for being here.
>
> We had hoped to conduct today a thorough examination of the Defense Department's V-22 Osprey, an aircraft with a controversial past, a troubled present, and an uncertain future.
>
> However, the Defense Department has evidently decided to stonewall our investigation. On May 5, 2009, I wrote to Secretary of Defense Gates to request information on the Osprey, including copies of two reports on the performance of the Osprey in Iraq, called "Lessons and Observations." I also requested a list of all V-22 Ospreys acquired by the Defense Department, including their current locations and flight status.
>
> However, to this date, the Defense Department has failed to provide this information, despite repeated reminders from the Committee. This is simply unacceptable.
>
> General Trautman, I want you to carry this message back to the Pentagon: We will pursue this investigation even harder than we have so far. We will not be slow-rolled. We will not be ignored.
>
> I intend to conduct a full investigation of the Osprey, not just an investigation of the information that you want me to see. We hope you will provide it voluntarily, but if you do not, we will compel your compliance.
>
> To ensure a thorough investigation and to allow the Defense Department additional time to provide us with these records, we will continue this hearing in two weeks and I am asking the witnesses to return to present their testimony at that time. This hearing is now adjourned, to be resumed in two weeks at the call of the chair.

(continued...)

Appendix A. Legislative Activity in 2011

(...continued)

Thank you.

(Source: Text of opening statement of Representative Edolphus Towns, as posted on the committee's website. http://oversight house.gov/documents/20090521101314.pdf. The listed witnesses for the hearing were Mr. Mike Sullivan, Director of Acquisition and Sourcing Management, Government Accountability Office; Mr. Dakota L. Wood, Senior Fellow, Center for Strategic and Budgetary Assessments; Lieutenant General George Trautman, Deputy Commandant for Aviation, U.S. Marine Corps; and Lieutenant Col Karsten Heckl, Commander, Marine Medium Tiltrotor Squadron 162 (VMM-162). See also Christopher J. Castelli, "Committee Accuses DOD of Stonewalling on V-22 Documents, Ends Hearing Abruptly," *InsideDefense.com (DefenseAlert – Daily News)*, May 21, 2009; and Geoff Fein, "House Oversight Committee Chair Claims DoD 'Stonewalling' V-22 Investigation," *Defense Daily*, May 22, 2009: 2-3.

On May 22, 2009, it was reported that:

> The Pentagon is denying the House Oversight and Government Reform Committee's accusations that it is stonewalling lawmakers' requests for information about the V-22 Osprey.

> "The Department of Defense coordination process is highly complex," Pentagon spokeswoman Cheryl Irwin told InsideDefense.com. "We are diligently working to fulfill this request and will have it to the proper officials in order that the hearing process can continue."

> House Oversight Committee Chairman Edolphus Towns (D-NY) yesterday accused the Pentagon of stonewalling his request for V-22 documents and vented his displeasure by abruptly ending a hearing after mere minutes, telling a three-star Marine Corps general to return in two weeks.

> Towns said the panel had hoped to conduct a "thorough examination" of the V-22 program, which he said has "a controversial past, a troubled present, and an uncertain future." But the Defense Department has "evidently decided to stonewall our investigation," he complained.

> The panel's ranking Republican, Rep. Darrell Issa (CA), also complained about DOD's failure to provide the documents, stressing the committee needs such information well in advance of any hearing. In a statement released later, he faulted a "bureaucratic failure of the Office of the Secretary of Defense," not the Marine Corps.

> After about three minutes, Towns ended the hearing. He said it would be continued in two weeks to give DOD additional time to provide the records. The witnesses were not invited to speak during the brief hearing nor did they attempt to do so. After the hearing, Lt. Gen. George Trautman, the Marine Corps' top aviation official and one of a handful of witnesses who had been scheduled to testify, declined to speak to reporters.

> Later that day, Marine Corps spokesman Maj. Eric Dent told InsideDefense.com the service understands Towns' decision to postpone the hearing. But the Marine Corps was disappointed "that we did not get the opportunity to discuss with the committee the Osprey's remarkable performance in Iraq over the past 19 months," he added. The V-22 program has nothing to hide, according to Dent.

> "As we were today, we remain prepared to discuss every aspect of the Osprey program with Congress," he said. "We are fully committed to openness and transparency; in fact, we've been working hand-in-hand with the Government Accountability Office for the past year in its own review of the Osprey program."...

> Dent insisted the Marine Corps is making a good-faith effort to address the request.

> "We forwarded, at the committee's request, more than 500 pages of maintenance records, after-action reports, and additional information on every MV-22 we have," he said. "Essentially, this was an aircraft-by-aircraft daily record of location and maintenance discrepancies. Collecting this information was a monumental task. Although we cannot speak to why the committee did not receive the information the Marine Corps prepared, we must emphasize that we have a process by which information, including classified material that was asked for by the committee, must be vetted before being released."

(Christopher J. Castelli, "Pentagon Denies Accusations of Stonewalling Congress on V-22," *InsideDefense.com [DefenseAlert – Daily News]*, May 22, 2009.)

Proposal to Eliminate V-22 Funding in FY2011

On February 15, 2011, the House voted 326 to 105 against a proposed cut to the FY2011 V-22 budget (H.Amdt. 13 to H.R. 1, Full-Year Continuing Appropriations Act, 2011, roll call vote 43).

The amendment proposed to eliminate FY2011 funding for the V-22 by reducing the amount for Aircraft Procurement, Navy by $22 million and for Aircraft Procurement, Air Force by $393 million.

Proponents cited cost overruns and argued that the V-22 did not meet operational requirements, citing the 2009 GAO report. Opponents noted the V-22's performance in Iraq and Afghanistan and highlighted advances made since the aircraft's development period.

FY2011 Funding Request for Procurement of V-22s

MV-22s

Procurement funding for MV-22s is in the Aircraft Procurement, Navy (APN) appropriation account, which funds the procurement of Navy and Marine Corps aircraft.

The Navy estimates the procurement cost of the 30 MV-22s requested for FY2010 at $2,267.6 million, or an average of about $75.6 million each. These 30 aircraft have received $146.6 million in prior-year advance procurement funding, leaving another $2,121.0 million requested in the APN account for FY2011 budget to complete their cost. The APN account also requests $81.9 million in advance procurement funding for V-22s that the Navy wants to procure in future fiscal years, bringing the total FY2011 APN funding request for procurement and advance procurement of MV-22s to $2,202.1 million.

CV-22s

Procurement funding for CV-22s is divided between the Aircraft Procurement, Air Force (APAF) appropriation account and the USSOCOM portion of the Procurement, Defense-Wide (PDW) appropriation account.

The Air Force estimates the APAF-funded portion of the procurement cost of the five CV-22s requested for FY2011 at $415.1 million, or an average of about $83.0 million in APAF funding for each. These five aircraft have received $22.1 million in prior-year APAF advance procurement funding, leaving another $393.1 million requested in the APAF account for FY2011 to complete the APAF-funded portion of their cost. The APAF account also requests $13.6 million in advance procurement funding for CV-22s that the Air Force wants to procure in future fiscal years, bringing the total FY2010 APAF funding request for procurement and advance procurement of CV-22s to $406.7 million.

Appendix B. V-22 Program History

This appendix provides additional discussion of the history of the V-22 program.

May 2009 GAO Report

A May 2009 GAO report provided the following summary of the history of the V-22 program:

> The Osprey program was started in December 1981 to satisfy mission needs for the Army, Navy, and Air Force. Originally spearheaded by the Army, the program was transferred to the Navy in 1982 when the Army withdrew from the program citing affordability issues. The program was approved for full-scale development in 1986, and the first aircraft was flown in 1989. A month after the first flight, the Secretary of Defense stopped requesting funds for the program due to affordability concerns. In December 1989, DOD directed the Navy to terminate all V-22 contracts because, according to DOD, the V-22 was not affordable when compared to helicopter alternatives, and production ceased. Congress disagreed with this decision, however, and continued to fund the project. Following a crash in 1991 and a fatal crash in 1992 that resulted in seven deaths, in October of 1992 the Navy ordered development to continue and awarded a contract to a Bell Helicopter Textron and Boeing Helicopters joint venture (Bell-Boeing) to begin producing production-representative aircraft.
>
> In 1994, the Navy chartered a medium lift replacement COEA, which reaffirmed the decision to proceed with the V-22. It also provided an analytical basis for KPPs to be proposed for the system. This analysis defined the primary mission of a medium-lift replacement aircraft to be the transport of combat troops during sea-based assault operations and during combat operations ashore. Secondary missions included transporting supplies and equipment during assault and other combat operations as well as supporting Marine Expeditionary Unit (MEU) special operation forces, casualty and noncombatant evacuation operations, tactical recovery of aircraft and personnel operations, combat search and rescue operations, and mobile forward area refueling and re-arming operations. These original mission descriptions and aircraft employment were reaffirmed by the Marine Corps in 2003 and again in 2007. The existing medium-lift aircraft fleet needed to be replaced due to inventory shortfalls and reduced aircraft reliability, availability, and maintainability—needs accentuated by the increasing age and limited capabilities of its current fleet of helicopters.
>
> The analysis concluded that the V-22 should be the Marine Corps' choice. The analysis considered a number of helicopter candidates—including the CH-46E and CH-53D—and the V-22 tiltrotor—judging each candidate based on their performance characteristics and expected contribution to tactics and operations. A sensitivity analysis was conducted which measured candidate aircraft against specific performance parameters—including KPPs. The analysis used models to assess research and development, production or procurement, and operations and support cost and concluded that for non-assault missions, such as medical evacuation missions, the V-22 was the most effective option because of its greater speed, increased range, and ability to deploy in one-third the time of the alternative candidates. For assault missions, the analysis concluded the V-22 would build combat power in the form of troops and equipment most quickly, was more survivable, would maximize the arrival of forces and minimize casualties, and would halve helicopter losses. In terms of affordability, the analysis concluded that, holding V-22 and helicopter force sizes equal, the V-22 would be the most effective but at a higher cost. The analysis further noted that while the major factor in favor of the V-22 was its speed, at short distances greater speed offers little advantage.

Subsequently, Low-Rate Initial Production (LRIP) began with five aircraft in 1997, increasing to seven each year in 1998 and 1999. In 2000, the program undertook operational evaluation testing, the results of which led the Navy's operational testers to conclude that the MV-22 was operationally suitable for land-based operations and was operationally effective. Later evaluations resulted in testers concluding that the MV-22 would be operationally suitable on ships as well. Based on the same tests, DOD's independent operational testers concluded that the MV-22 was operationally effective but not operationally suitable, due in part to reliability concerns. Despite the mixed test conclusions, a Program Decision Meeting was scheduled for December 2000 to determine whether the V-22 should progress beyond LRIP production and into full-rate production. Following two fatal crashes that occurred in 2000 and resulted in 23 deaths, the last one occurring just before the full-rate production decision, the V-22 was grounded and, rather than proceeding to full-rate production, the program was directed to continue research and development at a minimum sustaining production rate of 11 aircraft per year.

Before the V-22 resumed flight tests, modifications were made to requirements and design changes were made to the aircraft to correct safety concerns and problems. The aircraft nacelles were redesigned to preclude line chafing; a robust software qualification facility was built; and Vortex Ring State, a dangerous aerodynamic phenomenon that all rotor wing aircraft are subject to and was reported to have contributed to one of the fatal V-22 crashes in 2000, was further investigated. Requirements for landings in helicopter mode in which engine power had failed ("autorotation") and nuclear, chemical and biological weapons protection among others were eliminated, and some KPPs were modified, prior to conducting a second round of operational testing with modified aircraft in June 2005. Testers then recommended that the aircraft be declared operationally effective and suitable for military use. The Defense Acquisition Board approved it for military use as well as full-rate production in September 2005. DOD is procuring the V-22 in blocks. Block A is a training configuration, while later blocks are being procured and fielded as the operational configurations. Tables 1 and 2 provide a summary of the upgrades to be incorporated in each block configuration.[33]

Additional Discussion[34]

Early Development

The first of six MV-22 prototypes was flown in the helicopter mode on March 19, 1989, and as a fixed-wing airplane on September 14, 1989. Prototype aircraft numbers three and four successfully completed the Osprey's first Sea Trials on the USS Wasp (LHD-1) in December 1990.

The fifth prototype crashed on June 11, 1991, on its first flight, because of incorrect wiring in a flight-control system; the fourth prototype crashed on July 20, 1992, while landing at Quantico Marine Corps Air Station, VA, killing seven people and destroying the aircraft. This accident was caused by a fire resulting from hydraulic component failures and design problems in the engine nacelles.[35]

---

[33] Government Accountability Office, *Defense Acquisitions[:] Assessments Needed to Address V-22 Aircraft Operational and Cost Concerns to Define Future Investments*, GAO 09-482, May 2009, pp. 7-9.

[34] The discussion in this section is retained from earlier versions of this CRS report.

[35] Former Secretary of Defense Cheney tried to terminate the program in 1989-92, but Congress continued to provide funds for development of the V-22. The George H. Bush Administration's FY1990 budget requested no funds for the
(continued...)

Flight tests were resumed in August 1993 after changes were incorporated in the prototypes. Flight testing of four full-scale development V-22s began in early 1997 when the first pre-production V-22 was delivered to the Naval Air Warfare Test Center in Patuxent River, MD. The first Engineering and Manufacturing Development (EMD) Flight took place on February 5, 1997. The first of four low-rate initial production (LRIP) aircraft, ordered on April 28, 1997, was delivered on May 27, 1999. Osprey number 10 completed the program's second Sea Trials, this time from the USS Saipan (LHA-2), in January 1999.

Operational evaluation (OPEVAL) testing of the MV-22 began in October 1999 and concluded in August 2000. On October 13, 2000, the Department of the Navy announced that the MV-22 had been judged operationally effective and suitable for land-based operations. On November 15, 2000, the Marine Corps announced that the Osprey had successfully completed sea trials and had been deemed operationally effective and suitable for both land and sea-based operations.

Successfully completing OPEVAL should have cleared the way for full rate production. This decision was to have been made in December 2000, but was postponed indefinitely, because of a mixed report from DOD's director of operational test and evaluation, and two fatal accidents.

On April 8, 2000, another Osprey crashed near Tucson, AZ, during an exercise simulating a noncombatant evacuation operation. All four crew members and 15 passengers died in the crash. An investigation of the accident found that the pilot was descending in excess of the recommended flight envelope, which may have caused the aircraft to experience an environmental condition known as "power settling" or "vortex ring state." According to Lieutenant General Fred McCorkle, the pilot was descending more than 1,000 feet per minute. The recommended descent rate is 800 feet per minute. Following a two-month suspension of flight testing, the Osprey recommenced OPEVAL in June 2000, with pilots flying a slightly tighter flight envelope. A July 27, 2000, report by the Marine Corps Judge Advocate General (JAG) (which had access to all non-privileged information from the safety investigation) confirmed that a combination of "human factors" caused the crash.

> This mishap appears not to be the result of any design, material or maintenance factor specific to tilt ... rotors. Its primary cause, that of an MV-22 entering a Vortex Ring State (Power Settling) and/or blade stall condition is not peculiar to tilt rotors. The contributing factors to the mishap, a steep approach with a high rate of descent and slow airspeed, poor aircrew coordination and diminished situational awareness are also not particular to tilt rotors.[36]

A DOD Inspector General study concluded that the V-22 would not successfully demonstrate 23 major operational effectiveness and suitability requirements prior to the December 2000 OPEVAL Milestone III decision to enter full rate production in June 2001.[37] The Marine Corps agreed with DOD's assessment of the deficiencies, but said that they had been aware of these

---

(...continued)

program. In submitting that budget to Congress on April 25, 1989, Defense Secretary Cheney told the House Armed Services Committee that he "could not justify spending the amount of money ... proposed ... when we were just getting ready to move into procurement on the V-22 to perform a very narrow mission that I think can be performed ... by using helicopters instead of the V-22."

[36] V-22 JAGMAN Executive Summary, United States Marine Corps, Division of Public Affairs, July 27, 2000, p.1.

[37] *Audit Report: V-22 Osprey Joint Advanced Vertical Aircraft.* Report No. D-2000-174. Office of the Inspector General. Department of Defense. August 15, 2000.

deficiencies before the beginning of OPEVAL. Furthermore, the Marine Corps said that they had an approved plan designed to resolve the deficiencies prior to the Milestone III decision.

On November 17, 2000, DOD's Director of Operational Test and Evaluation issued a mixed report on the Osprey; saying although "operationally effective" the V-22 was not "operationally suitable, primarily because of reliability, maintainability, availability, human factors and interoperability issues." The report recommended that more research should be conducted into the V-22's susceptibility to the vortex ring state blamed for the April 8, 2000, crash.

On December 11, 2000, an MV-22 Osprey crashed near Jacksonville, NC, killing all four Marines on board. This was the fourth Osprey crash since 1991 and the third lethal accident. The aircraft's pilot, Lieutenant Colonel Keith M. Sweeney was the program's most experienced pilot and was in line to command the first squadron of Ospreys. The aircraft's copilot, Major Michael Murphy was second only to Sweeney in flying time on the Osprey.[38] The Marine Corps grounded the Osprey fleet pending a mishap board investigation. On April 5, 2001, the Marine Corps reported that the crash was caused by a burst hydraulic line in one of the Osprey's two engine casings, and a software malfunction that caused the aircraft to accelerate and decelerate unpredictably and violently when the pilots tried to compensate for the hydraulic failure.[39] The Marine Corps report called for a redesign of both the hydraulics and software systems involved.[40]

Maintenance and Parts Falsifications

In December 2000, an anonymous letter was mailed to the media by someone claiming to be a mechanic in the Osprey program. The letter claimed that V-22 maintenance records had been falsified for two years, at the explicit direction of the squadron commander. Enclosed in the letter was an audio tape that the letter's author claimed was a surreptitious recording of the squadron commander directing maintenance personnel to lie about the aircraft until the V-22 LRIP decision was made. On January 20, 2001, it was reported that the V-22 squadron commander admitted to falsifying maintenance records. The Marine Corps subsequently relieved him of command and reassigned him to a different position. At a May 1, 2001, hearing, members of the Senate Armed Services Committee expressed their concern that false data might impede DOD's ability to accurately evaluate the V-22 program and identify problem areas and potential improvements. The Department of Defense's Inspector General (IG) conducted an investigation. On September 15, 2001, it was reported that three Marines were found guilty of misconduct and two were reprimanded for their actions.

In June 2005, a U.S. grand jury indicted a company that had supplied titanium tubing for the V-22 program. The indictment charged the company with falsely certifying the quality of the tubes. The V-22 test program was halted for 11 days in 2003 because of faulty tubes. Replacing deficient tubes cost the V-22 program $4 million. Navy officials do not believe that these deficient tubes caused fatal mishaps.[41]

---

[38] James Dao, "Marines Ground Osprey Fleet After Crash Kills Four," *New York Times*, December 12, 2000.

[39] An un-redacted version of JAG investigation into the April 2000 V-22 crash indicates that investigators found three "noteworthy" maintenance "areas of concern", including the Osprey's hydraulics system. A Naval Safety Center presentation to the Blue Ribbon Panel brought to light several previously unreported maintenance problems—including hydraulics failures—that caused engine fires or other problems during the Osprey's operational testing.

[40] Mary Pat Flaherty, "Osprey Crash Blamed on Leak, Software," *Washington Post*, April 6, 2001.

[41] Louise Story. "Maker of Tubes for Osprey Aircraft is Indicted." *New York Times.* June 8, 2005. Christopher J. (continued...)

Reviews and Restructuring

On April 19, 2001, a Blue Ribbon panel formed by then-Secretary of Defense William Cohen to review all aspects of the V-22 program, reported its findings and recommendations.[42] These findings and recommendations were also discussed during congressional testimony on May 1, 2001. The panel recommended that the program continue, albeit in a restructured format. The panel concluded that there were numerous problems with the V-22 program—including safety, training and reliability problems—but nothing inherently flawed in basic tilt-rotor technology. Because of numerous safety, training, and reliability problems, the V-22 was not maintainable, or ready for operational use.

The panel recommended cutting production to the "bare minimum" while an array of tests were carried out to fix a long list of problems they identified with hardware, software, and performance. Cutting near-term production was hoped to free up funds to pay for fixes and modifications. Once the changes had been made and the aircraft was ready for operational use, the Panel suggested that V-22 out-year purchases could be made in large lots using multi-year contracts to lower acquisition costs. Program officials estimated that the minimal sustainable production rate is 12 aircraft per year, which would be less than half the Ospreys once planned for FY2002.[43] In P.L. 107-107 Sec.123, congressional authorizers codified the Blue Ribbon Panel's recommendation to produce V-22s at the minimum sustainable rate until the Secretary of Defense can certify that the Osprey is safe, reliable, maintainable, and operationally effective.

DOD appeared to take managerial and budgetary steps to incorporate the Blue Ribbon Panel's recommendations. For example, DOD's FY2001 supplemental funding request asked for a reduction of $475 million in procurement and an increase of $80 million in R&D funds. The additional R&D funding was to be used to support initial redesign and testing efforts to address deficiencies, logistics, flight test, and flight test support for V-22 aircraft. The reduction in procurement funding reflected the need to reduce production to the minimum rate while the aircraft design changes are being developed and tested.

Secretary of Defense Rumsfeld's FY2002 budget amendment, unveiled June 27, 2001, included a request for the procurement of 12 Ospreys. DOD comptroller Dov Zakheim and Marine Corps Commandant General James Jones both stated that the procurement of 12 aircraft in FY2002 would allow them to sustain the V-22 subcontractor base while simultaneously addressing the Osprey program's needs.[44] V-22s were procured at a rate of 11 per year from FY2002 to FY2006.

Following the Blue Ribbon panel's recommendations, former DOD Under Secretary for Acquisition Edward "Pete" Aldridge assumed acquisition authority for the V-22 program. Under Secretary Aldridge changed the V-22 program's status from an ACAT 1C program—which gives the Department of the Navy the highest required authority for production decisions—to an ACAT

(...continued)

Castelli. "Former Supplier of Hydraulic Tubing for V-22 Osprey Faces Indictment." *Inside the Navy.* June 13, 2005.

[42] This panel was chaired by retired Marine General John R. Dailey and included retired Air Force General James B. Davis, Norman Augustine, and MIT professor Eugene Covert.

[43] Adam Hebert, "Minimal Sustainable Rate Will Dramatically Cut Near-Term V-22 Buys," *Inside the* Air Force, April 20, 2001.

[44] DOD News Briefing, Wed. June 27, 2001, 1:30PM and Kerry Gildea, "New V-22 Plan Sustains Lower Tier Contractors, Jones Reports," Defense Daily, May 15, 2001.

1D program. Under the latter category, the Defense Acquisition Board (DAB) would decide if and when the program is ready to enter full rate production.[45]

A NASA-led review of the V-22 program, released November 6, 2001, concluded that there were no known aero-mechanical phenomena that would stop the tilt-rotor aircraft's development and deployment. The study focused on several aero-mechanics issues, including Vortex Ring State, power problems, auto-rotation, and hover performance.[46]

In a December 21, 2001, memo to the Secretaries of the Air Force and the Navy, and the Commander, Special Operations Command, Under Secretary of Defense Aldridge gave his authorization for the V-22 to resume flight testing in the April 2002 time frame. Secretary Aldridge expressed support for range, speed, and survivability goals of the V-22. He noted, however that the program still had numerous technical challenges to overcome, and emphasized that the V-22 must demonstrate that "1) it can meet the needs of the warfighter better than any other alternative, 2) it can be made to be reliable, safe, and operationally suitable, and 3) it is worth its costs in contributing to the combat capability of U.S. forces." Secretary Aldridge approved the flight test program under the condition that the production rate be slowed to the minimum sustaining level, that it be comprehensive and rigorous, and that the restructured program is fully funded in accordance with current estimates.[47] Under Secretary Aldridge estimated that the V-22 would require at least two years of flight testing before DOD could conclude that the aircraft is safe, effective, and "worth the cost."[48]

Mechanical adjustments slowed the V-22 test schedule, and the MV-22 took its first test flight on May 29, 2002. The Air Force CV-22 resumed flight tests on September 11, 2002. Flight tests were designed to explore both technical and operational concerns. Technical concerns include flight control software and the reliability and robustness of hydraulic lines. Operational concerns explored included whether the Osprey is too prone to Vortex Ring State to make it a safe or effective aircraft, whether this potential problem is further exacerbated by multiple Osprey's flying in formation, and how well the V-22 handles at sea.[49]

The principal differences between the aircraft that were grounded in 2000 and the aircraft that began testing 17 months later (called "Block A" aircraft) are re-routed hydraulic lines, and an improved caution and warning system.[50] Technical glitches were experienced during tests. Hydraulic failures, for example, continued during the reinstated flight test program, once on August 4, 2003, (due to a mis-installed clamp) and again on September 5, 2003. In June 2004 a V-22 was forced twice to make an emergency landing. During one landing, the aircraft suffered a "Class B" mishap (one causing between $200,000 and $1 million in damage).[51] An investigation

---

[45] "Navy Loses Osprey Authority," Washington Post, May 22, 2001, and Hunter Keeter, "Aldridge Maneuvers V-22 Acquisition Authority Away from Navy," *Defense Daily*, May 22, 2001, and Linda de France, "V-22 Osprey Production Authority Transferred from Navy to DoD," *Aerospace* Daily, May 22, 2001.

[46] Christopher Castelli, " NASA Review Panel Endorses Resumption of V-22 Flight Tests," InsideDefense.com,. November 14, 2001.

[47] "Text: Aldridge Memo on V-22," *Inside the Navy*, January 7, 2002.

[48] Tony Capaccio, "Textron-Boeing V-22 Needs Two years of Testing, Aldridge Says," *Bloomberg.com*, October 16, 2001.

[49] Thomas Ricks, "V-22 Osprey to Face Make or Break Tests," *Washington Post*, December 25, 2002, p. 14.

[50]Jefferson Morris, "Pilot: Resumption of V-22 Testing To Be Treated Like First Flight," *Aerospace Daily,* April 29, 2002.

[51] Christopher Castelli. "Navy Convenes Mishap Board to Investigate Latest V-22 Incident." *Inside the Navy*. July 5, (continued...)

---

revealed that the V-22 suffered from widespread problems with an engine component that required replacement every 100 flight hours.[52]

In conjunction with resuming flight testing, the Navy Department modified certain V-22 requirements. For instance, the V-22 is no longer required to land in helicopter mode without power (also known as "autorotation"), protection from nuclear, chemical and biological weapons has been eliminated. The V-22 is no longer required to have an "air combat maneuvering" capability; instead it must demonstrate "defensive maneuvering." Also, the requirement that troops be able to use a rope or rope ladder to exit the cabin at low altitudes has been eliminated.[53] Also concurrent with the resumption of V-22 flight testing, DOD began an in-depth study of alternatives to pursue in case the aircraft does not pass muster. Options reportedly include purchasing the S-92, or upgrading CH-53, or EH101 helicopters.[54]

After one calendar year and 466 hours of flight testing, DOD reviewed the Osprey's progress. On May 15, 2003, Thomas Christie, DOD's Director of Operational Test and Evaluation (DOT&E), graded Bell-Boeing's improvements to the Osprey's hydraulics as "reasonable and appropriate" and "effective."[55] Christie also at that time approved of the testing that had been completed and was satisfied with what had been learned about the V-22's susceptibility to Vortex Ring State. On May 20, 2003, the Defense Acquisition Board also reviewed the program and approved of the flight test program's progress.

Marine Corps officials recommended increasing the production rate in FY2006 from the minimum sustainable rate of 11 to 20 aircraft. However, in an August 8, 2003, memorandum, Under Secretary of Defense for Acquisition Michael Wynne announced that this acceleration "presents more risk than I am willing to accept." Instead, Wynne restructured the planned procurement, reducing the FY2006 purchase to 11 aircraft. "For subsequent years' procurement planning, production rates should increase by about 50% per year for a total of 152 aircraft through FY09," according to the August 8 memo. Wynne directed that the savings resulting from the reduced procurement (estimated at $231 million) be invested in improving the V-22's interoperability, by funding the Joint Tactical Radio System, Link 16 and Variable Message Format communication. Wynne also directed that a multi-year procurement (MYP) of the V-22 be accelerated. While some suggest that this restructuring will more quickly deliver high-quality aircraft to the Marines and Special Operations Forces, others fear that slowing procurement will inevitably raise the platform's cost.

In December 2004 the V-22 budget and schedule were restructured again. Program Budget Decision 753 (PBD-753) cut 22 aircraft from the V-22's production schedule and $1.3 billion from the budget between FY2006 and FY2009.

On June 18, 2005, the MV-22 program completed its second round of operational evaluation (OPEVAL) flight. The test program was marked by two emergency landings, a Class B mishap, a

---

(...continued)
2004.

[52] Christian Lowe. "V-22 Ospreys Require New Engine Component Every 100 Hours." *Navy Times.* July 16, 2004.

[53] Joseph Neff, "Eased Standards 'Fix' Osprey," *Raleigh News & Observer,* May 19, 2002, p.1.

[54] "Aldridge Makes Progress Check on MV-22 at NAS Patuxent River," *Defense Daily,* February 11, 2003.

[55] Tony Capaccio, "Boeing-Textron B-22 Gets Favorable Review From Pentagon Tester," *Bloomberg.com,* May 19, 2003.

small fire in an engine compartment, and problems with the prop-rotor gear box. However, Navy testers recommended that DOD declare the V-22 operationally suitable, and effective for military use. This recommendation was based, in part, on observations that the MV-22 had complied with the objectives of P.L. 107-107 Sec.123: hydraulic components and flight control software performed satisfactorily, the aircraft was reliable and maintainable, the MV-22 operated effectively when employed with other aircraft, and the aircraft's downwash did not inhibit ground operations.[56]

On September 28, 2005, the V-22 program passed a major milestone when the Defense Acquisition Board approved it for military use and full rate production.[57] The MV-22 continues testing to assess survivability and to develop tactics. The CV-22 is in developmental test and evaluation. The program continues to experience technical and operational challenges, and mishaps. For example, an inadvertent takeoff in March 2006 caused wing and engine damage in excess of $1 million. An engine component has been replaced because its failure in flight has caused seven unexpected flight terminations. In October 2005, a V-22 experienced engine damage during flight due to icing. An engine compressor failure during the V-22's first overseas deployment (July 2006) forced the aircraft to make a precautionary landing before reaching its destination. An engine fire on December 7, 2006, caused more than $1 million to repair, and the Marine Corps grounded all of its V-22s in February 2007 after it was found that a faulty computer chip could cause the aircraft to lose control during flight.

Author Contact Information

Jeremiah Gertler
Specialist in Military Aviation
jgertler@crs.loc.gov, 7-5107

---

[56] "Letter of Observation in Support of MV-22 Program Compliance with Section 123 of the National Defense Authorization Act for Fiscal Year 2002." Commander, Operational Test and Evaluation Force. Department of the Navy. February 18, 2005.

[57] Andy Pasztor. "Pentagon Clears Full Production for Osprey Aircraft." *Wall Street Journal.* September 29, 2005.

---

www.ingramcontent.com/pod-product-compliance
Lightning Source LLC
Chambersburg PA
CBHW081416170526
45166CB00010B/3370